Air-Release, Air/Vacuum, and Combination Air Valves

AWWA MANUAL M51

First Edition

FOUNDED 1881

American Water Works Association

MANUAL OF WATER SUPPLY PRACTICES—M51, First Edition

Air-Release, Air/Vacuum, and Combination Air Valves

Project manager and copy editor: Melissa Christensen
Production editor: Carol Stearns

Library of Congress Cataloging-in-Publication Data
Air-release, air/vacuum, and combination air valves.-- 1st ed.
p. cm. -- (AWWA manual ; M51)
Includes bibliographical references and index.
ISBN 1-58321-152-7
1. Water-pipes--Valves. 2. Air valves. I. American Water Works Association. II. Series.

TD491 .A49 no. M51
628.1 s--dc21
[628.1'5] 2001053999

Printed in the United States of America

American Water Works Association
6666 West Quincy Avenue
Denver, CO 80235

ISBN 1-58321-152-7

Printed on recycled paper

Contents

This page intentionally blank.

Figures

This page intentionally blank.

Tables

This page intentionally blank.

Preface

This manual is a guide for selecting, sizing, locating, and installing air valves in water applications. It is a discussion of recommended practice, not an American Water Works Association (AWWA) standard. It provides guidance on generally available methods and capacity information. Questions about specific situations or applicability of specific valves should be directed to the manufacturer or supplier.

Information contained in this manual is useful for operators, technicians, and engineers for gaining a basic understanding of the use and application of air valves. There are many special water pipeline applications that are beyond the scope of the methodology given in this manual and may require special tools such as computer programs for analysis of hydraulic transients. The valve capacity information is generic information. Actual capacity charts of the intended manufacturer's valve should be consulted before making the final selection of valve size and options. The manual provides information only on the air valve types listed in AWWA Standard C512, latest edition, including the following:

- Air-release valve
- Air/vacuum valve
- Combination air valve

Wastewater air valves, vacuum breakers, slow-closing air valves, and throttling devices are only introduced in this manual. Other sources of information should be consulted for the use and application of these devices.

This manual refers to AWWA standards, which are available for purchase from the AWWA Bookstore by calling (800) 926-7337 or online at <www.awwa.org/bookstore>.

Manufacturers graciously provided valve illustrations and other documentation. AWWA does not endorse any manufacturer's products, and the names of the manufacturers have been removed from the material provided.

Metrication Note: Valve sizes are listed in their current US designation, i.e., nominal pipe sizes in inches. To obtain an approximate metric equivalent, use a conversion factor of 25.4 mm per inch.

This page intentionally blank.

Acknowledgments

The AWWA Standards Committee on Water Works Air-Release Valves, which developed this manual, had the following personnel at the time of approval.

M.E. Wollam, Chair, Montgomery Watson, Pasadena, Calif.

J.V. Ballun, Val-Matic Valve & Manufacturing Corp., Elmhurst, Ill.
J.H. Bambei Jr., Denver Water, Denver, Colo.
J.J. Cusack Jr., Parsons, Brinckerhoff, Quade & Douglas, Boston, Mass.
R. Dilorenzo, Cla-Val Automatic Control Valves, Wauconda, Ill.
Gary Fuller, Henningson/Durham/Richardson, Denver, Colo.
N.E. Gronlund, East Bay Municipal Utilities District, Oakland, Calif.
Keith Hall, APCO Valve & Primer Corporation, San Clemente, Calif.
J.E. Herold, Seattle Public Utilities, Seattle, Wash.
W.A. Hunt, Consulting Engineer, Bozeman, Mont.
P.O. Landon, Val-Matic Valve & Manufacturing Corp., Elmhurst, Ill.
J.E. Lescovich, G.A. Industries Inc., Cranberry Township, Pa.
B.J. Lewis, Multiplex Manufacturing, Berwick, Pa.
W.L. Meinholz, Alvord, Burdick & Howson, Chicago, Ill.
W.J. Nicholl, G.A. Industries Inc., Cranberry Township, Pa.
J.L. Patton, Black & Veatch LLP, Kansas City, Mo.
J. Radtke, Aqua-Dynamic Systems, Wilkes-Barre, Pa.
M.I. Schwartz, Loudon County Sanitation Authority, Leesburg, Va.
R.J. Wahanik, Hystras, Wyomissing, Pa.
J.S. Wailes, Standards Engineer Liaison, AWWA, Denver, Colo.
R.A. Ward, Dufresne & Associates, PC, Windsor, Vt.
J.G. Yannotta, Los Angeles Water & Power, Los Angeles, Calif.

Others who provided valuable help with this manual include the following:

G.J. Tupac, G.J. Tupac & Associates Inc., Pittsburgh, Pa.
M.B. Horsley, Black & Veatch LLP, Kansas City, Mo.

This page intentionally blank.

Chapter 1

Introduction

Air valves are hydromechanical devices designed to automatically release or admit air during the filling, draining, or operation of a water pipeline or system. The safe operation and efficiency of a pipeline are dependent on the continual removal of air from the pipeline. This chapter includes an explanation of the effects of air and the sources of air in a pipeline.

OCCURRENCE AND EFFECT OF AIR IN PIPELINES

Water contains at least two percent dissolved air by volume in standard conditions (14.7 psia and 60° F)(Dean, 1992) but can contain more, depending on the water pressure and temperature within the pipeline. Henry's law states that "the amount of gas dissolved in a solution is directly proportional to the pressure of the gas above the solution" (Zumdahl, 1997). Therefore, when water is pressurized, its capacity to hold air is greatly magnified. The bubbling in soft drinks occurs after they are opened because the pressure over the fluid is reduced, and the excess carbon dioxide gas rapidly escapes. In a water system, a similar condition may occur at the consumer's tap when excess air comes out of solution. Once out of solution, air will not readily return to solution and will collect in pockets at high points along the pipeline.

Air comes out of solution in a pipeline because of low-pressure zones created by partially open valves, cascading flow in a partially filled pipe, variations in flow velocity caused by changing pipe diameters and slopes, and changes in pipeline elevation.

An air pocket may reduce the flow of water in a pipeline by reducing the cross-sectional flow area of the pipeline and may, if the volume of the air pocket is sufficient, completely air bind the pipeline and stop the flow of water (Karassik, 2001).

Generally, the velocity of the flow of water past an enlarging air pocket is sufficient to prevent complete air binding of the pipeline by carrying part of the air pocket downstream to collect at another high point. Although the flow velocity of water flow may prevent the pipeline from complete air binding, air pockets will increase head loss in the pipeline (Edmunds, 1979). Additional head loss in a pipeline decreases the flow of water and increases power consumption required to pump the water. Air pockets in pipelines are difficult to detect and will reduce the pipeline system's overall efficiency.

Air pockets may also contribute to water hammer problems, pipeline breaks, pipeline noise, and pipeline corrosion, and can cause erratic operation of control valves, meters, and equipment.

SOURCES OF AIR ENTRY INTO PIPELINES

In addition to air coming out of solution, air may enter pipelines at leaky joints where the pressure within the pipeline falls below atmospheric pressure. These conditions exist in the vortex at the pump suction, at pump glands where negative pressure occurs, and all locations where the pipeline lies above the hydraulic grade line.

Air may enter pipelines through air/vacuum and combination air valves following complete pump shutdown, through the orifices of air-release valves installed in pipeline locations where the pipeline pressure is less than atmospheric, and through pump suction pipes that are not properly designed to prevent vortexing. Finally, vertical turbine and well pumps start with air in the pump column, which may pass by the check valve and flow into the pipeline.

REFERENCES

Dean, John A. 1992. *Lange's Handbook of Chemistry*. New York; McGraw Hill

Edmunds, Robert C. 1979. "Air Binding In Pipes," *Journal AWWA*. May, pp. 272–277.

Karassik, Igor J., et al. 2001. *Pump Handbook*, McGraw Hill, New York.

"Theory, Application, and Sizing of Air Valves," 1997. Val-Matic Valve & Mfg. Corp.

Zumdahl, Steven S. *Chemistry*, third edition.

Chapter 2

Types of Air Valves

This chapter describes the three basic types of air valves used in the water industry that are included in AWWA C512, latest edition, "Standard for Air-Release, Air/Vacuum, and Combination Air Valves for Waterworks Service."

AIR-RELEASE VALVES

Air-release valves, also called small orifice valves, are designed to automatically release small pockets of accumulated air from a pipeline while the system operates under pressure exceeding atmospheric pressure. A typical air-release valve mechanism is shown in Figure 2-1. Air-release valves are characterized by outlet orifices, which are much smaller than the inlet connection or pipe size. Orifice sizes are generally between $^1/_{16}$ in. (1.6 mm) and 1 in. (25 mm) in diameter, while the inlet connections can range from $^1/_2$ in. (13 mm) to 6 in. (150 mm) in diameter.

When received, the valve is normally open and will vent air through the orifice. As water enters the valve, the float rises, closing the orifice. When air, which has accumulated in the piping system, enters the valve, it replaces the water, causing the float to drop and allowing the air to vent through the orifice. An air-release valve designed with the proper float weight and leverage mechanism will allow the valve to open at any pressure up to the maximum working pressure of the valve.

AIR/VACUUM VALVES

Air/vacuum valves, also called large orifice valves, are designed to exhaust large quantities of air automatically during pipeline filling and to admit large quantities of air automatically when the internal pressure drops below atmospheric pressure. The negative pressure may be caused by column separation, pipeline draining, pump failure, or a break in the pipeline. A typical air/vacuum valve is shown in Figure 2-2. Air/vacuum valves are characterized by orifices between $^1/_2$ in. (13 mm) and 20 in. (500 mm) diameter that match the nominal inlet size of the valve when built in accordance with AWWA C512. As a pipeline fills with water, the air in the pipeline must be expelled smoothly and uniformly to minimize pressure surges. Likewise, after a power failure or as a pipeline drains, air must be admitted to the pipeline to prevent the formation of a vacuum, which may collapse some pipelines or cause surges in the system.

The operation of an air/vacuum valve is similar to the air-release valve except that the orifice diameter is considerably larger and will not open under pressure. An air/vacuum valve is normally open and is designed to vent large quantities of air through the orifice. As water enters the valve during filling of the system, the float will rise closing the orifice. Air/vacuum valves once closed WILL NOT REOPEN TO VENT AIR while the pipeline is operating under pressure exceeding atmospheric pressure or if water is present.

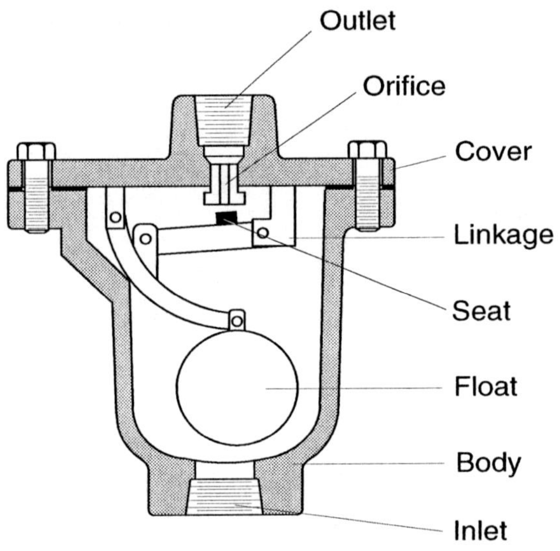

Figure 2-1 Air-release valve

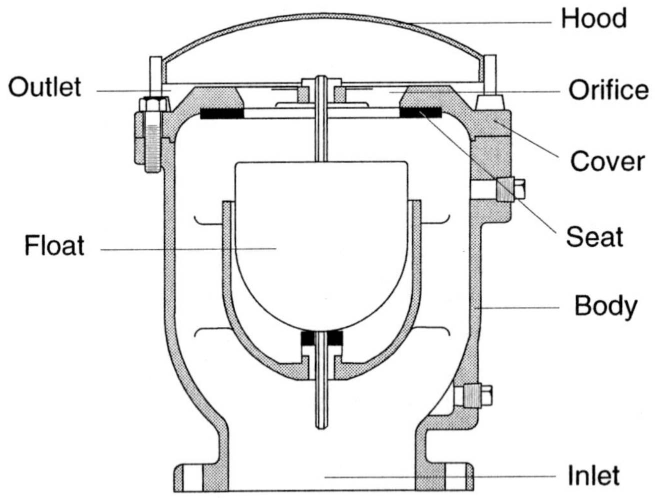

Figure 2-2 Air/vacuum valve

COMBINATION AIR VALVES

Combination air valves are designed to perform the same function as air/vacuum valves but, in addition, they will automatically release small pockets of air from the pipeline while under pressure like an air-release valve. Combination air valves can be supplied in a single-body configuration or a dual-body configuration as shown in Figure 2-3.

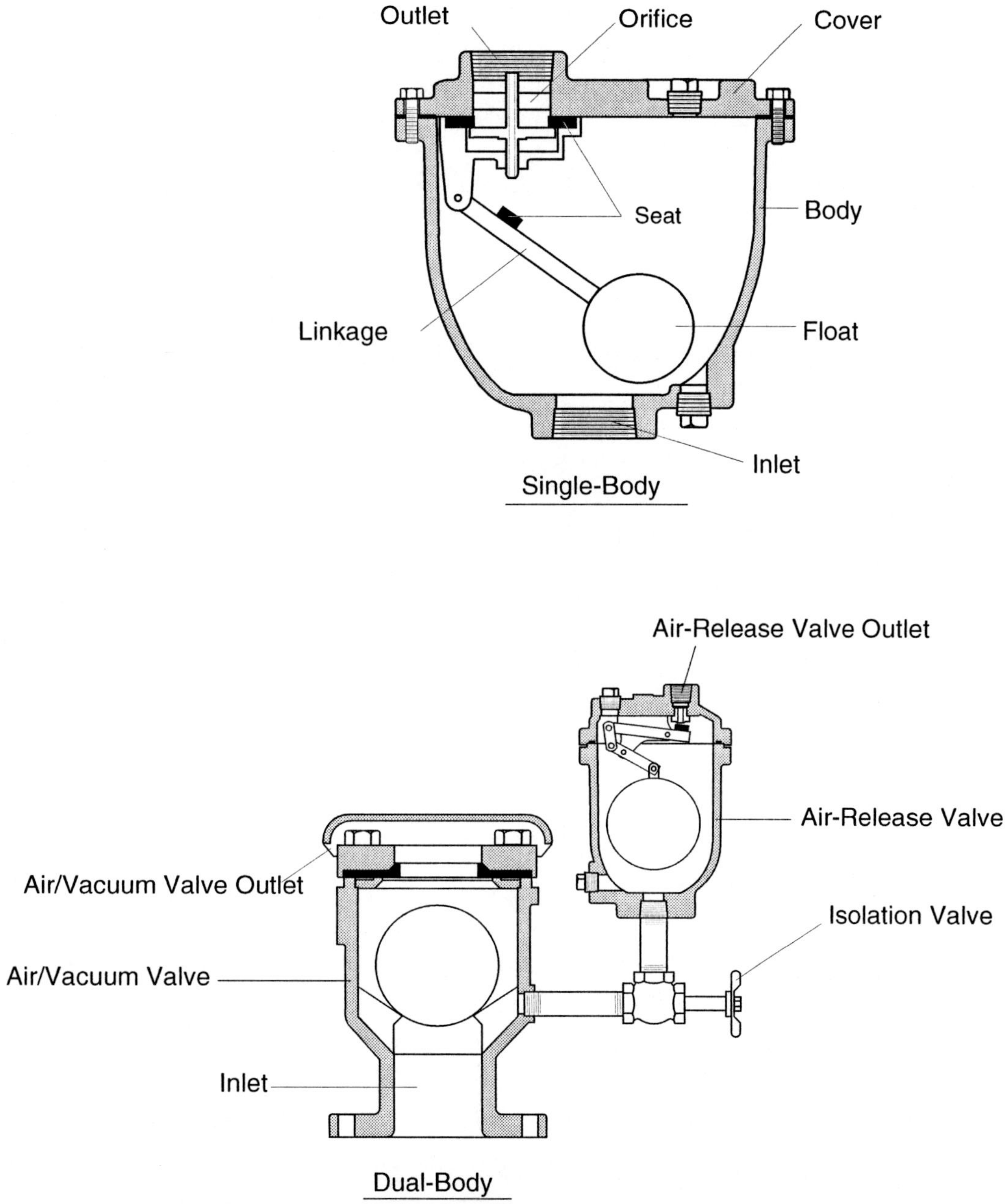

Figure 2-3 Single-body and dual-body comination air valves

This page intentionally blank.

Chapter 3

Locating Air Valves Along a Pipeline

This chapter addresses the location of air valves along a pipeline for the elimination of air pockets, which could potentially cause air binding, and for pipeline drainage. The information in this chapter is intended to apply generally to transmission pipelines but may also apply to other situations. This manual does not address the location or use of air valves for downsurge and column separation control, which should be considered for some systems.

PIPELINE LOCATIONS

The proper location of air-release, air/vacuum, and combination air valves is as important as the proper size of the valve. An improper location can render the valve ineffective. The following guidelines are recommended for the general location and corresponding types of air valves. However, there may be other locations where valves may be deemed necessary.

A sample pipeline profile illustrating typical valve locations is shown in Figure 3-1. The horizontal axis is the running length of the pipeline, usually expressed in station points. Station points are often expressed in hundreds of feet, such as 145+32, which is equivalent to 14,532 feet. The vertical axis is the elevation of the profile stations relative to a specified horizontal datum.

Air valves are typically used in transmission pipelines where raw water is being transported to a treatment plant or where finished water is transported to a distribution system, or similar applications. Air valves may not be needed on smaller piping in distribution system piping grids where hydrants and service connections provide means for venting trapped air. Hydrants may also provide a means for venting pipelines for drainage. Experience has shown that hydrants and service connections can provide sufficient removal of air in terms of both performance and cost.

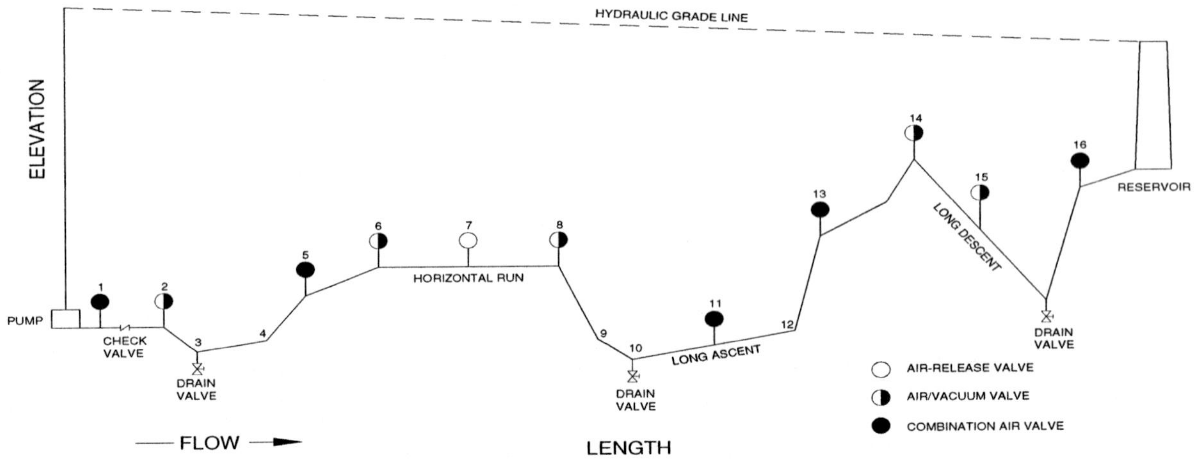

No.	Description	Recommended Types	No.	Description	Recommended Types
1	Pump Discharge	Air/Vac	9	Decr. Downslope	No Valve Required
2	Incr. Downslope	Combination	10	Low Point	No Valve Required
3	Low Point	No Valve Required	11	Long Ascent	Air/Vac or Combination
4	Incr. Upslope	No Valve Required	12	Incr. Upslope	No Valve Required
5	Decr. Upslope	Air/Vac or Combination	13	Decr. Upslope	Air/Vac or Combination
6	Beg. Horiz.	Combination	14	High Point	Combination
7	Horizontal	Air-Rel or Combination	15	Long Descent	Air-Rel or Combination
8	End Horiz.	Combination	16	Decr. Upslope	Air/Vac or Combination

Figure 3-1 Sample pipeline profile illustrating typical valve locations

Suggested locations and types

Air valves should be installed at the following locations.

- High Points. Combination air valves should be installed at pipeline high points to provide venting while the pipeline is filling, during normal operation of the pipeline, and for air inflow and vacuum protection while the pipe is draining. A high point is defined by the hydraulic gradient and is considered the upper end of any pipe segment that slopes up to the hydraulic gradient or runs parallel to it.

- Mainline Valves (not illustrated in Figure 3-1). Air/vacuum valves or combination air valves can be used on the draining side of mainline valves to facilitate draining of the pipeline.
- Increased Downslope. A combination air valve should be considered at abrupt increases in downslope.
- Decreased Upslope. An air/vacuum valve or a combination air valve should be considered at abrupt decreases in upslope.
- Long Ascents. An air/vacuum valve or combination air valve should be considered at intervals of $^1/_4$ mile (400 m) to $^1/_2$ mile (800 m) along ascending sections of pipelines.
- Long Descents. An air-release valve or combination air valve should be considered at intervals of $^1/_4$ mi (400 m) to $^1/_2$ mi (800 m) along descending sections of pipelines.
- Horizontal Runs. Combination air valves should be considered at the beginning and end of long horizontal sections, and air-release valves or combination air valves should be considered at intervals of $^1/_4$ mi (400 m) to $^1/_2$ mi (800 m) along horizontal sections of pipeline. It is difficult to evacuate air from a long horizontal pipeline at low-flow velocities.
- Venturi Meters (not illustrated in Figure 3-1). Air-release valves should be installed upstream of Venturi meters to eliminate measurement inaccuracies caused by trapped air.
- Deep Well and Vertical Turbine Pumps. Air/vacuum valves should be installed on the discharge side of deep well and vertical turbine pumps to remove the air in the well column during pump startup and to allow air to reenter the line after pump shutdown. Air valves mounted on these types of pumps may require special consideration in selection because of the violent changes in flow rate during pump cycling. Air-release valves are often used with time-delayed, power-actuated check valves to release the air in the pump column slowly under full pump pressure (Val-Matic Valve, 1997).
- Siphons (not illustrated in Figure 3-1). To maintain a siphon on a section of pipeline that extends above the hydraulic gradient and that constantly runs under negative pressure, install an air-release valve on the high point of the siphon to vent the air. However, the air-release valve must be equipped with a vacuum check device on the outlet to prevent admitting air into the pipeline. For systems requiring more venting capacity, a similar approach can be accomplished with an air/vacuum valve with vacuum check device on the outlet.

When reverse flow is undesirable after pump stoppage, a specialized air/vacuum antisiphon valve can be used. An antisiphon valve is designed to vent air during start-up, close tight during flowing conditions, and open to break the siphon during reverse-flow conditions using a flow paddle.

REFERENCE

"Theory, Application, and Sizing of Air Valves," 1997. Val-Matic Valve & Mfg. Corp.

This page intentionally blank.

Chapter 4

Design of Valve Orifice Size

It is important to select the proper size valve orifice for the specific location along the pipeline. This chapter provides a common methodology used in the water industry based on formulas and data tables. Numeric examples are provided for clarity. For specific sizing of valves, refer to manufacturers' charts, graphs, and formulas; the figures presented in this chapter only demonstrate the methods used.

SIZING FOR RELEASING AIR UNDER PRESSURE

The orifice size for releasing air under pressure is generally between $^1/_{16}$ in. (1.6 mm) and 1 in. (25 mm) in diameter; however, the size of the valve inlet connection can range from $^1/_2$ in. (13 mm) to 6 in. (150 mm) in diameter with the smaller orifices found in the smaller-sized inlet port and higher-pressure valves.

There is no definitive method for determining the amount of air that may need to be vented from a given pipeline. This is because of the difficulty in predicting the quantity of air that will enter the pipeline or come out of solution as the pressure varies along the pipeline. A common method is to provide sufficient capacity to release two percent of the flow of water in terms of air at standard conditions (Lescovich, 1972). This method is based on the 2 percent solubility of air in water at standard conditions. The air is vented through the orifice of the air-release valve at the pipeline working pressure at that valve location.

Because of the high pressures involved, the applicable flow equation for air flow through an orifice is based on compressible adiabatic flow where there is no heat transfer to the air. Sonic flow will occur when discharging air at a pressure exceeding 1.9 times the outlet pressure. Assuming that the outlet pressure is atmospheric pressure (14.7 psia [101 kPa (absolute)]), then any inlet pressure exceeding 1.9 times 14.7, or 28 psia (13 psig [90 kPa (gauge)]), will produce sonic flow (ASME, 1971). At sonic flow, the air velocity is limited to the speed of sound, thereby causing a restriction to the air discharge at higher pressures.

For the purpose of generating the tables and graphs in Table 4-1 and Figure 4-1, sonic flow and a discharge coefficient of 0.7 was assumed. A discharge coefficient of 0.7 is an approximation and falls between a smooth flow nozzle and a square-edged orifice. The actual discharge coefficient of the valve and piping will be different. Therefore, the capacity charts of valve suppliers should be consulted before selecting the final valve size.

The working pressure of an air-release valve is calculated with reference to the maximum hydraulic grade line at the valve and not the pump discharge head. The working differential pressure at the air-release valve location is the difference between the valve elevation and the maximum hydraulic gradient elevation at the valve.

The following method may be used to approximate the orifice size required in an air-release valve. It is important to verify with the supplier that the valve will operate with the required orifice diameter at the expected maximum line pressure. Valve capacity information is presented in both tabular and graphic form to suit the preference of the user. A flow formula is also provided to calculate the capacity of varying orifice diameters at any pressure condition.

ORIFICE SIZING METHOD FOR RELEASING AIR

Step 1. Divide the pipeline flow rate in gallons per minute (gpm) by 7.48 to obtain flow in cubic feet per minute (cfm).

Step 2. Multiply the flow in cfm from step 1 by 0.02 to determine the required air venting volume, as two percent of the pipeline flow in standard cubic feet per minute (scfm). *Standard* refers to air at the conditions of 60° F and 0 psi.

Step 3. Determine the working pressure at the valve by subtracting the valve elevation from the hydraulic grade elevation. Express the pressure in pounds per square inch (psi). If the elevations are in feet, multiply by 0.433 to obtain psi.

Step 4. Refer to Table 4-1 or Figure 4-1 and select the orifice diameter that provides the required capacity from step 2 at the pressure from step 3. Consult the available orifice sizes from suppliers and select the valve that meets both the capacity and pressure requirements of the application.

Table 4-1 Air capacity table of air-release valve orifices (C_d = 0.7)

Pressure (*psi*)	Orifice Diameter, *In.*									
	1/16	3/32	1/8	3/16	1/4	5/16	3/8	7/16	1/2	1
25	1.6	3.5	6.3	14.2	25.2	39.4	56.7	77.1	100	400
50	2.6	5.8	10.3	23.1	41.0	64.1	92.3	126	164	656
75	3.6	8.0	14.2	32.0	56.9	88.9	128	174	228	910
100	4.5	10.2	18.2	40.9	72.8	114	164	223	291	1,160
125	5.5	12.5	22.2	49.8	88.6	138	199	271	354	1,420
150	6.5	14.7	26.1	58.8	104	163	235	320	418	1,610
175	7.5	16.9	30.1	67.7	120	188	271	369	481	1,920
200	8.5	19.2	34.1	76.6	136	213	306	417	545	2,180
225	9.5	21.4	38.0	85.5	152	238	342	466	608	2,430
250	10.5	23.6	42.0	94.5	168	262	378	514	672	2,690
275	11.5	25.8	45.9	103	184	287	414	563	735	2,940
300	12.5	28.1	49.9	112	200	312	449	611	799	3,200

NOTE: Metric conversions—in. × 25.4 = mm, cfm × 0.4719 = L/sec, psi × 6.89476 = kPa.

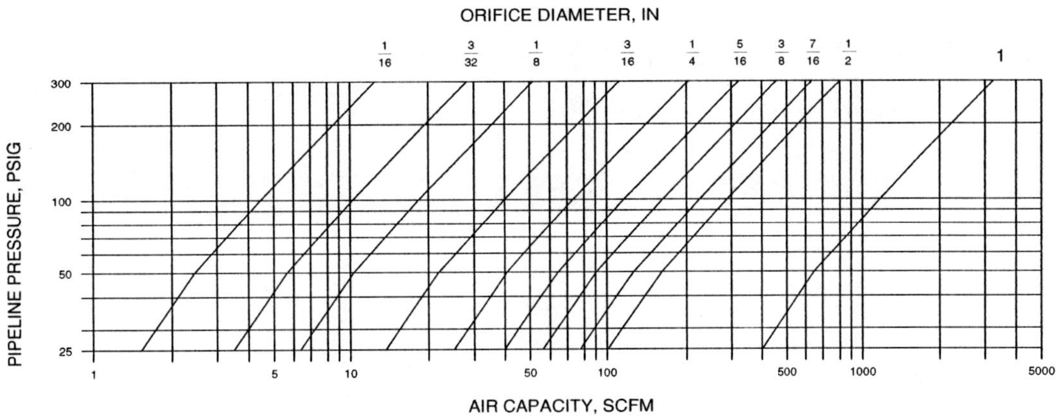

Figure 4-1 Discharge of air through small orifice, cfm

Example

A pipeline with a flow rate of 10,500 gpm requires an air-release valve at a location with a valve elevation of 600 feet and a hydraulic grade line elevation of 831 feet.

1. 10,500 gpm /7.48 = 1,404 cfm
2. 1,404 × 0.02 = 28 scfm
3. (831 – 600) × 0.433 = 100 psi
4. Select 3/16 in. orifice from Table 4-1 that provides 40.9 scfm at 100 psi.

The capacity information shown in Table 4-1 and Figure 4-1 is based on the compressible adiabatic flow equation and sonic flow (Technical Paper No. 410, 1982).

$$Q = 678 Y d^2 C_d \sqrt{\Delta P P_1 / (T S_g)} \qquad \text{(Eq 4-1)}$$

Where:

Q	=	flow rate, scfm
Y	=	expansion factor, 0.71 for air flow (Technical Paper No. 410, 1982)
d^2	=	orifice diameter, in^2.
C_d	=	coefficient of discharge, 0.7
ΔP	=	differential pressure, 0.47 P_1 (for sonic flow)
P_1	=	inlet pressure, psia (pipeline pressure + 14.7 psi) (assumes sea level atmospheric pressure of 14.7 psia; pressure will vary with altitude)
T	=	inlet temperature, 520 degrees Rankine
S_g	=	specific gravity, 1.0 (for air)

For subsonic conditions where pipeline pressures are generally less than 13 psig (90 kPa [gauge]):

$$Q = 14.77\, d^2 \left[\Delta P (P + 14.7)\right]^{1/2} \qquad \text{(Eq 4-2)}$$

Where:

P	=	pipeline pressure, psig

SIZING FOR PIPELINE FILLING

For the initial filling of a pipeline, air should be vented at the same volumetric rate as the pipeline is being filled. In many cases, one pump is turned on until the line is full. The recommended procedure, however, is to fill the pipeline at a gradual rate to prevent surges in the line. A suggested filling rate is about 1 ft/sec (0.3 m/sec). For more information, see the discussion of water hammer in chapter 5.

The volumetric rate of air from initial filling is vented to atmosphere at a typical differential pressure of 2 psi (13.8 kPa). Valves equipped with antislam or slow-closing devices may be sized with a differential pressure of 5 psi (34.5 kPa). The following method may be used to approximate the orifice size required for pipeline filling. Generic tables, graphs, and formulas are provided to suit the preference of the user.

The applicable flow equation is based on compressible adiabatic flow through a short nozzle or tube where there is no heat transfer to the air. Also, it is assumed that the valve is at sea level and a temperature of 60°F (15.5°C). At high altitudes or extreme temperatures, equations of a more general nature should be used. For the purpose of generating the tables and graphs in Table 4-2 and Figure 4-2, a discharge coefficient of 0.7 is used. A discharge coefficient of 0.7 is an approximation and falls between a smooth flow nozzle and a square-edged orifice. Therefore, capacity charts of valve suppliers should be consulted before selecting the final valve size.

Orifice sizing method for pipeline filling

(Assumes air valve is at sea level and 60°F [15.5°C]).

Step 1. Calculate the venting flow rate in scfm using:

$$Q = q\,(.134\ \text{ft}^3/\text{gal})\ \frac{(\Delta P + 14.7\ \text{psi})}{(14.7\ \text{psi})} \qquad \text{(Eq 4-3)}$$

Where:

Q = flow rate, scfm
q = fill rate, gpm
ΔP = differential pressure, 2 psi

Step 2. Refer to Table 4-2 or Figure 4-2 and select the orifice diameter that provides the required flow at the selected venting pressure.

Example

A 66-in. pipeline will fill at a flow rate of 10,500 gpm (1 ft/sec), and the air valve will vent the air at a pressure of 2 psi.

1. Q = (10,500) (.134) (2.0 + 14.7 / 14.7) = 1,598 scfm
2. Referring to Table 4-2 and Figure 4-2, at 2 psi, select a 4 in. orifice that will vent 1,780 scfm.

Table 4-2 Air discharge table of large orifices (C_d = 0.7, T = 60°F, sea level)

Differential Pressure (*psi*)	Orifice Diameter, *In.* 1	2	3	4	6	8	10	12	14	16	18	20
1.0	79	317	712	1,270	2,850	5,070	7,910	11,400	15,500	20,200	25,600	31,700
1.5	97	387	870	1,550	3,480	6,190	9,670	14,000	18,900	24,700	31,300	38,600
2.0	111	445	1,000	1,780	4,010	7,120	11,100	16,000	21,800	28,500	36,100	44,500
2.5	124	497	1,120	1,990	4,470	7,950	12,400	17,900	24,300	31,800	40,200	49,600
3.0	136	543	1,220	2,170	4,890	8,690	13,600	19,500	26,600	34,700	44,000	54,300
3.5	146	585	1,320	2,340	5,270	9,370	14,600	21,100	28,700	37,500	47,400	58,500
4.0	156	625	1,410	2,500	5,620	10,000	15,600	22,500	30,600	40,000	50,600	62,500
4.5	165	662	1,490	2,650	5,960	10,600	16,500	23,800	32,400	42,300	53,600	66,200
5.0	174	697	1,570	2,790	6,270	11,100	17,400	25,100	34,100	44,600	56,400	69,700

NOTE: Metric conversions—in. × 25.4 = mm, cfm × 0.4719 = L/sec, psi × 6.89476 = kPa.

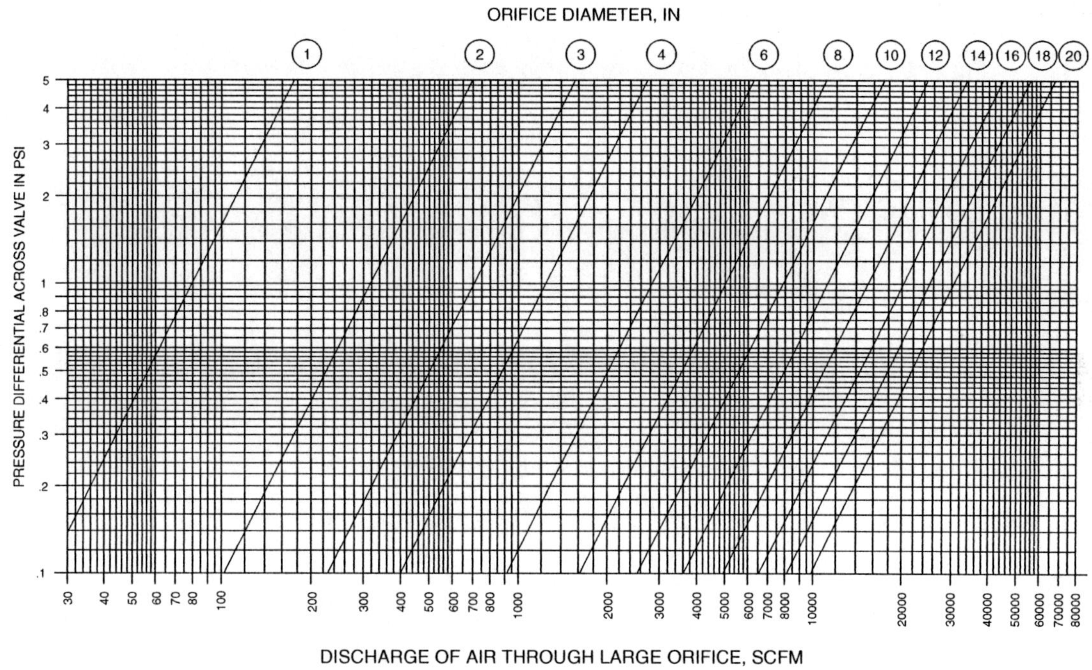

Figure 4-2 Air discharge graph of large orifices (C_d = 0.7).

SIZING FOR PIPELINE DRAINING

When it is necessary to drain a pipeline for repairs, the pipeline should be drained at a controlled rate of about 1–2 ft/sec (0.3–0.6 m/sec) to minimize pressure transients. An air valve at the high point adjacent to the draining location must be sized to admit air at the same volumetric rate as the pipeline being drained.

SIZING FOR GRAVITY FLOW

A power failure or line break may result in a sudden change in the flow velocity because of column separation and gravity flow. The gravity flow may result in excessive vacuum conditions occurring at the adjacent high points. Most small and medium-size pipes commonly used in the water industry can withstand a complete vacuum; however, because of low stiffness, large-diameter pipelines may collapse from negative internal pressures. Therefore, sizing air valves for gravity flow conditions is important to maintaining the integrity of the pipeline. Air valves at high points should be sized to allow the inflow of air to minimize negative pressures in the pipeline and prevent possible damage to pump seals, equipment, or the pipeline itself.

When sizing an air valve orifice for gravity flow, the pipe slope will determine the volume of air required to prevent excessive vacuum. An appropriate air valve should be provided at the nearest high point with the orifice sized to allow the required inflow of air to replace the water in the pipeline. Figure 4-3 illustrates the required inflow of air required for various pipe sizes and slopes.

The orifice sizing of an air valve for inflow is typically based on the lower of 5 psi (34 kPa) or the allowable negative pressure below atmospheric pressure for the pipeline with a suitable safety factor. Sonic flow will occur when the outlet-to-inlet pressure ratio (ASME, 1971) falls below 0.53. Knowing that the inlet pressure is atmospheric pressure (14.7 psia [101 kPa]), then any negative pipeline pressure below 7.8 psia (54 kPa (absolute)) or –7 psig (48 kPa) (vacuum) will produce sonic flow. Because the flow will be sonic and restricted, flow volume will not increase beyond –7 psi (48 kPa) differential.

If gravity flow occurs in a pipeline with a change in slope where the pipeline lower section has a steeper slope than the upper section, then an air/vacuum valve should be considered at the location where the pipeline changes slope. The gravity flow will be greater in the pipeline section with the steeper slope. The air/vacuum valve orifice should be sized so that the inflow of air at this point equals the difference in the two flow rates at the allowable negative pressure.

The applicable flow equation is based on compressible adiabatic flow through a short nozzle or tube where there is no heat transfer to the air and subsonic flow. For the purpose of estimating circular orifice sizes, a discharge coefficient, C_d, of 0.7 was used to generate Table 4-3 and Figure 4-4. A discharge coefficient of 0.7 is an approximation and falls between a smooth flow nozzle and a square-edged circular orifice. Capacity charts of valve suppliers should be consulted before selecting the final valve size.

Table 4-3 Air inflow table of large orifices (C_d = 0.7)

Differential Pressure (*psig*)	Orifice Diameter, *In.*											
	1	2	3	4	6	8	10	12	14	16	18	20
1.0	76	306	688	1,220	2,750	4,890	7,650	11,000	15,000	19,600	24,800	30,600
1.5	92	366	824	1,470	3,300	5,860	9,160	13,200	17,900	23,500	29,700	36,700
2.0	103	414	931	1,660	3,720	6,620	10,300	14,900	20,300	26,500	33,500	41,400
2.5	113	452	1,020	1,810	4,070	7,230	11,300	16,300	22,100	28,900	36,600	45,200
3.0	121	484	1,090	1,930	4,350	7,740	12,100	17,400	23,700	31,000	39,200	48,300
3.5	127	510	1,150	2,040	4,590	8,160	12,700	18,400	25,000	32,600	41,300	51,000
4.0	133	532	1,200	2,130	4,780	8,510	13,300	19,100	26,100	34,000	43,000	53,200
4.5	137	550	1,240	2,200	4,950	8,800	13,700	19,800	26,900	35,200	44,500	55,000
5.0	141	565	1,270	2,260	5,080	9,030	14,100	20,300	27,700	36,100	45,700	56,500

NOTE: Metric conversions—in. × 25.4 = mm, cfm × 0.4719 = L/sec, psi × 6.89476 = kPa.

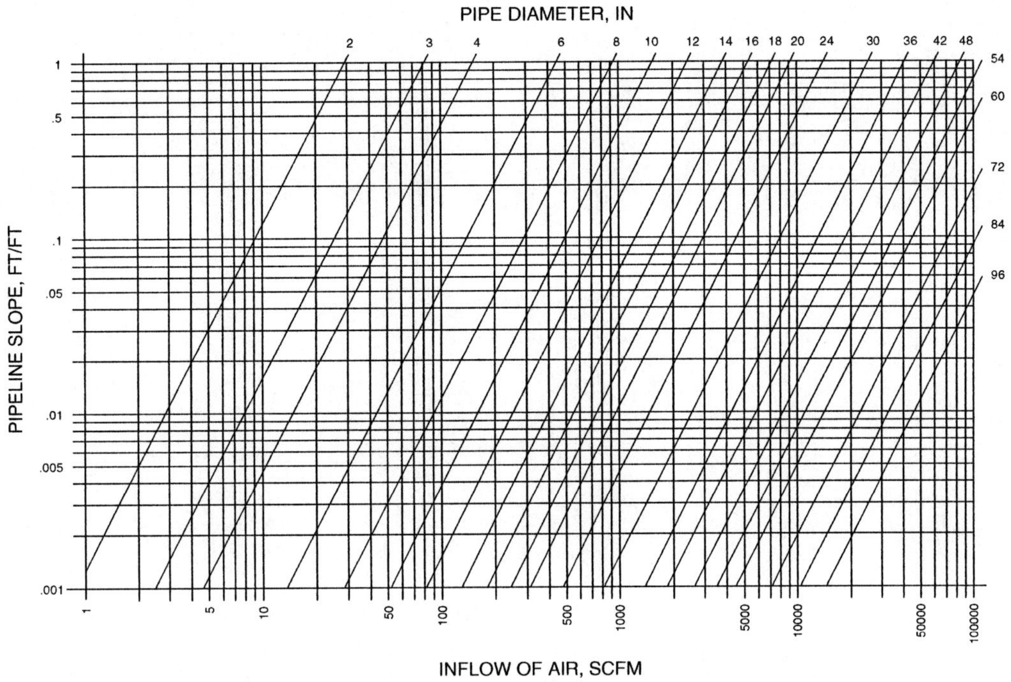

Figure 4-3 Inflow of air for gravity flow

Orifice sizing method for gravity flow

Step 1. Determine the allowable negative pressure for the pipeline with consideration of a reasonable safety factor. Consult the pipe manufacturer for the maximum recommended negative pressure. For low-stiffness, large-diameter steel pipe, the collapse pressure can be estimated by the general formula for collapse of thin-walled steel cylinders (AWWA M11, 1989). The formula is applicable to a pipe submerged or an aboveground environment. Pipes used in buried service with good soil compaction are not prone to vacuum collapse.

$$P_c = 66{,}000{,}000\,(t\,/d)^3 \qquad \text{(Eq 4-4)}$$

Where:

P_c = collapsing pressure, in.
t = pipe wall thickness, in.
d = mean diameter of pipe, in.

The allowable differential pressure for sizing is then found by the formula

$$\Delta P = P_c\,/\,SF \qquad \text{(Eq 4-5)}$$

Where:

ΔP = differential pressure, psi
SF = safety factor, dimensionless

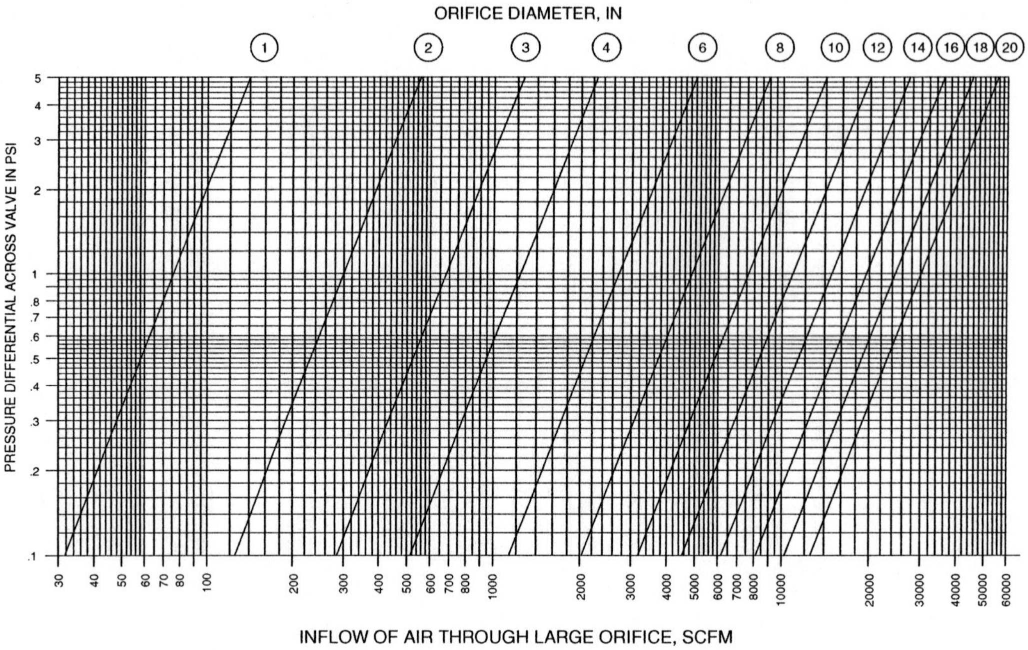

Figure 4-4 Air inflow graph of large orifices (C_d = 0.7)

The choice of safety factor (i.e., 3.0 or 4.0) is at the discretion of the pipeline designer. When the pipe is not subject to collapse, a differential pressure of 5.0 psi (34 kPa) is commonly used.

Step 2. Calculate the slope of the pipeline (*S*) as the change in elevation divided by horizontal distance (i.e., rise over run expressed in the same units, ft/ft).

Step 3. Determine the required air inflow in scfm from Figure 4-3 by matching the pipeline slope against the pipe diameter. For increases in downgrade and decreases in upgrade, compute the difference between the flows in the lower and upper sections of pipe. Flow rates can also be calculated using common flow formulas, such as Hazen-Williams, Manning, or the following formula:

$$Q = 0.0472C\sqrt{S\,ID^5} \qquad \text{(Eq 4-6)}$$

Where:

Q	=	flow rate, scfm
C	=	Chezy coefficient, 110 (iron), 120 (concrete), 130 (steel), 190 (polyvinyl chloride)
S	=	pipeline slope, ft/ft
ID	=	pipe inside diameter, in.

The coefficient, *C*, varies for different pipe roughness and is different from the C-factor associated with the Hazen-Williams formula.

Step 4. Refer to Table 4-3 or Figure 4-4 for selecting the orifice diameter that provides the required flow in scfm at the permissible differential pressure.

Example

Using the aboveground 24-in. ID by 1/8-in.-thick steel pipeline illustrated in Figure 4-5, calculate the minimum orifice diameter at stations 10+00 (assuming a line break at station 0+00), 25+00 (assuming a line break at station 20+00), and 40+00 (assuming a line break at station 20+00).

1. $d = ID + t$ = 24.000 in. + .125 in. = 24.125 in.
   P_c = 66,000,000 (.125 in. / 24.125 in.)^3 = 9.2 psi
   (from Equation 4-4)
   Assuming a safety factor of 4.0.
   ΔP = 9.2 psi/4.0 = 2.3 psi (from Equation 4-5)

2. S_1 = 40 ft/1,000 ft = 0.04
   S_2 = 40 ft/500 ft = 0.08
   S_3 = 20 ft/1,500 ft = 0.013

3. For S_1 = 0.04 and ID = 24, Figure 4-3 provides Q_1 = 3,000 scfm
   For S_2 = 0.08 and ID = 24, Figure 4-3 provides Q_2 = 4,050 scfm
   For S_3 = 0.013 and ID = 24, Figure 4-3 provides Q_3 = 1,900 scfm

   To size the orifice at station 25+00, Q_{25+00} = 4,050 – 1,900 = 2,150 scfm.

4. For station 10+00, use Table 4-3 and select a 6 in. orifice with an inflow capacity of about 4,000 scfm at 2.3 psi that exceeds Q_1 of 3,000 scfm.
   For station 25+00, use Table 4-3 and select a 6 in. orifice with an inflow capacity of about 4,000 scfm at 2.3 psi that exceeds Q_{25+00} of 2,150 scfm.
   For station 40+00, use Table 4-3 and select a 6 in. orifice with an inflow capacity of about 4,000 scfm at 2.3 psi that exceeds Q_3 of 1,900 scfm.

SIZING FOR SPECIAL APPLICATIONS

There are special situations requiring the application of air valves, such as the control of water column separation and the minimizing of subsequent pressure transients. Sizing of these valves is usually included in the transient analysis of a pipeline using a computer program and is beyond the scope of this manual.

In some cases, such as large-diameter pipes subject to collapse, the size of the air valve calculated in the section Sizing for Gravity Flow may be beyond the size range of standard manufactured valves. In these cases, it is suggested to install clusters of valves. Another alternative is to use a high-capacity vacuum breaker in combination with an air valve to provide the needed inflow capacity as shown in Figure 4-6.

The sizing of air valves for vertical turbine deep-well pump discharge service is highly dependent on the specific characteristics of the air valve and sometimes the pump. Therefore, these applications should be based on the published sizing recommendations of the air valve supplier. Deep-well pump applications are described further in chapter 5.

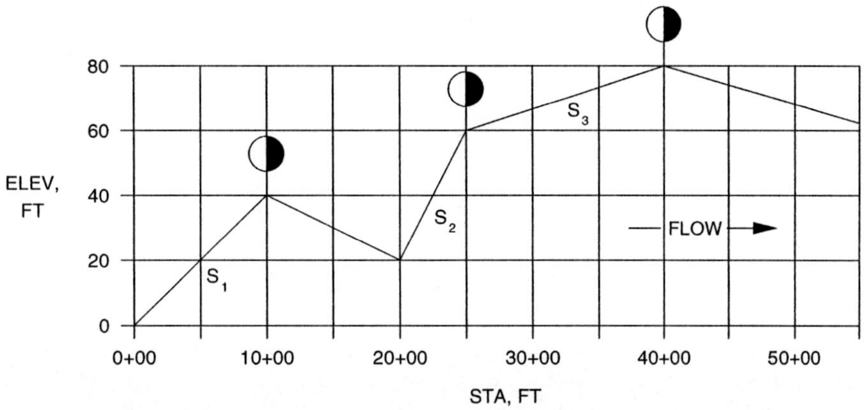

Figure 4-5 Example pipeline installation for gravity flow

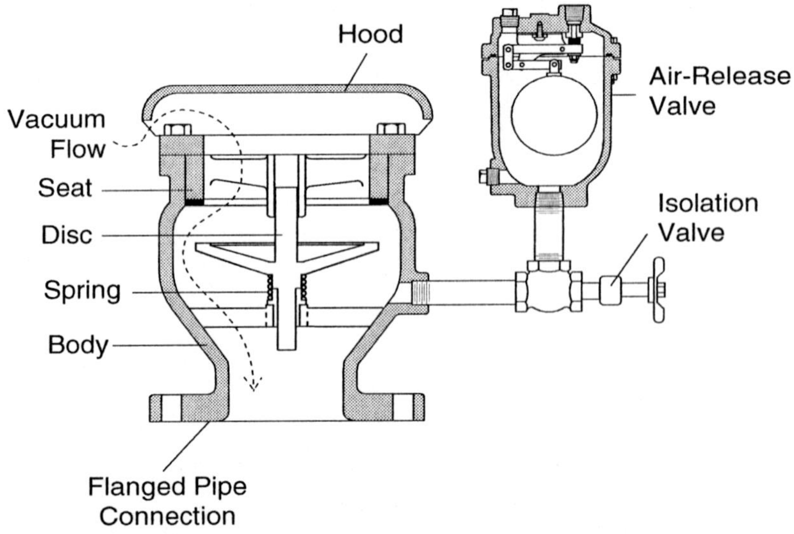

Figure 4-6 Vacuum breaker with air-release valve

AIR-RELEASE VALVE SELECTION

The following information is recommended for selecting the correct air-release valve for venting accumulated air during pipeline operation:

- Compliance with AWWA C512, latest edition
- Orifice size from the section Sizing for Releasing Air Under Pressure
- NPT inlet size
- Maximum working pressure of each valve
- Valve construction materials
- Type of installation (in-plant, in-vault, or outdoor)

The selection of a larger orifice or inlet size is acceptable as long as the maximum operating pressure is not exceeded.

For a given orifice size (e.g., $^1/_8$ in. [3 mm]), several inlet sizes may be available (e.g., $^1/_2$ in. [13 mm] to 6 in. [150 mm]). The inlet size should be as large as possible to maximize the air/water exchange in the pipeline connection. Also, the pipeline connection should never be less than the inlet size of the air-release valve.

The maximum working pressure of an air-release valve is related to the construction of the valve body and the mechanical advantage of the float leverage mechanism. The valve must have sufficient mechanical advantage to allow the weight of the float to pull the seal away from the orifice. Valves with large orifices (i.e., greater than $^1/_8$ in. [3 mm]) or high operating pressures (i.e., greater than 175 psi [1,206 kPa]) will usually employ a compound lever mechanism with a series of levers and pivot pins. It is important for the valve to have a maximum working pressure greater than the highest expected operating pressure at the specific valve location.

Typical options for air-release valves include special corrosion-resistant construction or a vacuum check on the valve outlet to prevent air from reentering the system during negative pressure conditions.

AIR/VACUUM VALVE SELECTION

The following information is recommended for selecting the correct air/vacuum valve for venting air during pipeline filling and admitting air during negative pressure conditions:

- Compliance with AWWA C512, latest edition
- Orifice size
- Inlet size and type of connection
- Maximum working pressure of each valve
- Valve construction materials
- Type of installation (in-plant, in-vault, or outdoor)
- Type of outlet connection (threaded, flanged, or hooded)

The orifice size must be sufficient to meet all of the requirements for

- Venting air during pipeline filling per section Sizing for Pipeline Filling
- Admitting air during pipeline draining per section Sizing for Pipeline Draining
- Admitting air during line break per section Sizing for Gravity Flow

Select a valve size that satisfies all three requirements. The inlet size for an air/vacuum valve generally matches the orifice size. Oversized air/vacuum valves should not be used where the potential for column separation exists or surges can result. The maximum pressure rating of the valve will influence the seat material in the valve. Typically, air/vacuum valves rated for high pressure (i.e., greater than 300 psi [2,068 kPa]) and large-diameter valves (i.e., greater than 14 in. [350 mm]) may be equipped with hard nonmetallic seats or stainless-steel seats containing o-ring seals.

Typical options for air/vacuum valves include special corrosion-resistant construction, screened hoods, and antislam or surge-check devices mounted on the inlet to reduce valve pressure surges.

COMBINATION AIR VALVE SELECTION

The following information is recommended for selecting the correct combination air valve for venting air during pipeline filling, admitting air during negative pressure conditions, and venting accumulated air during pipeline operation:

- Compliance with AWWA C512, latest edition
- Sizes of air-release and air/vacuum orifices
- Inlet size and type of connection
- Maximum working pressure of each valve
- Valve construction materials
- Type of installation (in-plant, in-vault, or outdoor)
- Type of outlet connection (threaded, flanged, or hooded)
- Body configuration (single-or dual-body)

The orifice size must be sufficient to meet all of the requirements for

- Venting accumulated air under pressure per section Sizing for Releasing Air Under Pressure
- Venting air during pipeline filling per section Sizing for Pipeline Filling
- Admitting air during pipeline draining per section Sizing for Pipeline Draining
- Admitting air during line break per section Sizing for Gravity Flow

Select a valve configuration that satisfies all four requirements.

Single-body configurations are typically more economical. They are more compact, less likely to freeze, and are tamper-resistant. Single-body configurations are limited in availability to a maximum size of 8 in. (200 mm). Dual-body configurations consist of an air-release valve piped to an air/vacuum valve. Many combinations and ranges of capacities are therefore available. Also, if the air-release valve is being serviced, the air/vacuum valve is still protecting the pipeline.

The inlet size for a combination air valve generally matches the orifice size of the air/vacuum orifice. Oversized combination air valves should not be used where the potential for column separation exists or surges can result. The maximum working pressure of the valve must also include the ability to vent air through the air-release orifice at the expected maximum pressure of the specific pipeline location.

Typical options for combination air valves include special corrosion-resistant construction, screened hoods, and antislam or slow-closing devices mounted on the inlet to reduce valve pressure surges.

REFERENCES

AWWA Standard for Air-Release, Air/Vacuum, and Combination Air Valves for Waterworks Service, AWWA Standard C512, latest edition, Denver, Colo.: AWWA.

Fluid Meters, Their Theory and Application. 1971. ASME, (6th edition).

Flow of Fluids. 1982. Technical Paper No. 410, Crane.

____. p. A-21 for k = 1.4, Beta = 0, Sonic Flow.

Giles, Ranald V. *Fluid Mechanics and Hydraulics,* (2nd Edition), McGraw Hill, New York., p. 160. Equation (2) was used to compute C based on friction factors of 0.021, 0.015, and 0.007.

Lescovich, J.E. Locating and Sizing Air-Release Valves, *Journal AWWA*, July, 1972.

Steel Pipe—A Guide for Design and Installation, AWWA M11, 3rd edition. 1989. Equation (4-3) (E = 30,000,000 psi, v = 0.30).

Chapter 5

Water Hammer Effects

Water hammer is a sudden rise in pressure resulting from rapid changes in flow velocity in pipelines and is also referred to as surge or transient pressure (AWWA M11, 1989). Water hammer is an extremely complex phenomenon requiring computer analysis; however, the use of general operating principles will minimize the effects of water hammer. This chapter presents some applications for air valves in systems where water hammer may occur.

AIR/VACUUM AND COMBINATION AIR VALVES

To minimize the effects of water hammer during filling of a pipeline, it is recommended that the pipeline filling velocity be maintained at 1 ft/sec (0.3 m/sec) or less. Properly designed air/vacuum or combination air valves will allow air to exhaust from the pipeline relatively unrestricted. However, when the last of the air escapes the pipeline, the air/vacuum or combination air valve may shut abruptly in response to the water reaching the valve float. The resulting collision between adjacent columns of water in the vicinity of the valve may cause a rapid deceleration of the water in the pipeline, resulting in a surge (Tullis, 1989). Air valves may be equipped with slow-closing devices to minimize the abrupt closing of the air/vacuum or combination air valves.

Air/vacuum or combination air valves are provided on pipelines to protect against pipe collapse under negative pressure conditions. These pipelines are especially prone to water hammer effects during the filling operations because the orifice diameter required for collapse criteria provides minimal air discharge regulation, especially at excessive filling rates. For these and other installations where large-diameter air valves are used, it is important to provide for strict control of the filling rate. This may require the throttling of the pump discharge flow rate or throttling the gravity supply flow rate during the filling operation. Generally, a filling rate that limits the pipeline velocities to 1 ft/sec (0.3 m/sec), is acceptable (Sanks, 1989).

AIR VALVES AT WELL PUMPS

Air/vacuum or combination air valves installed on the discharge of vertical turbine or deep-well pumps are subject to water hammer problems similar to those encountered in the filling of pipelines. Air needs to be vented from the pump column upon start-up. Otherwise, air may be delivered into the discharge pipeline when the check valve opens. Uncontrolled air exhaust and the abrupt closure of the air/vacuum valves on pump discharge applications can lead to serious pressure surges.

To minimize these water hammer effects, the pump discharge flow rate may be controlled at startup, or slow-closing devices or air-throttling devices may be incorporated into the air/vacuum valve design. These special devices, manufactured for vertical turbine and deep-well pump installations, generally regulate the exhaust rate and closure speed of the air/vacuum valve. It is important to note that the slow-closing and dampening devices are effective in suppressing water hammer only when placed near the pump. Figure 5-1 shows the proper location of an air/vacuum valve with slow-closing device.

Air-release valves can be used with time-delayed, power-actuated pump discharge control valves to release air in the pump column slowly under full pump pressure before the control check valve opens.

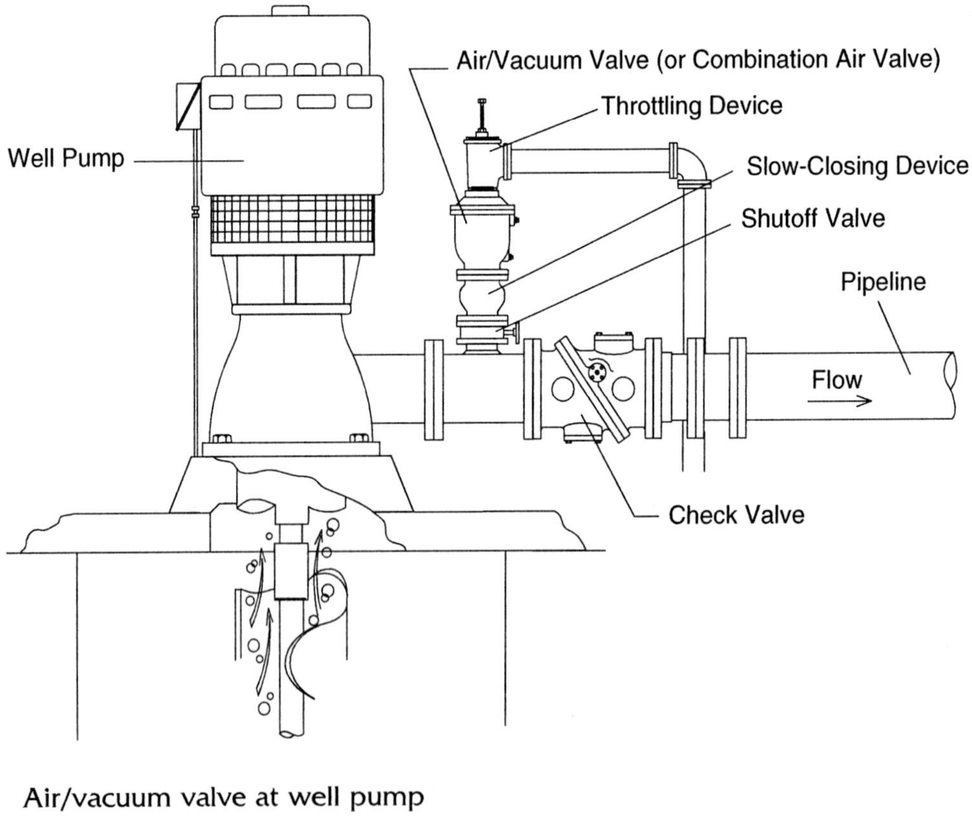

Figure 5-1 Air/vacuum valve at well pump

AIR VALVES ON PIPELINES

The presence of air in a transmission pipeline may reduce the conveyance capacity of the pipeline substantially. In underwater hammer conditions, entrapped air may magnify the surge problem. Trapped air can store energy and cause check-valve slamming. If air pockets become dislodged, water hammer can be caused when the air passes through restrictions, through partially open valves, or from one high point to another causing a change in velocity. Some general guidelines for minimizing the effects of air in a pipeline are as follows (Tullis, 1989):

1. Fill slowly, 1 ft/sec (0.3 m/sec) velocity.
2. Install properly sized air/vacuum or combination air valves so air is not released under high pressure during pipeline filling.
3. Lay the pipeline to a set grade and install air valves at high points. If the terrain is flat, install air valves at regular intervals.
4. Flush the system at moderate velocities, 2–4 ft/sec (0.6–1.2 m/sec), and low pressure to move the air to the air valves.
5. Install air valves upstream of control valves so air does not pass through modulating valves.
6. Use combination air valves wherever possible so that air flow is provided to accommodate filling, draining, and air accumulation.

Water hammer in pipelines can also be analyzed with special computer programs (Wood, latest edition). Water hammer software can provide immediate feedback of the effects of suggested air valve locations and sizes on system performance including:

- Valve size and location effects during pipeline filling
- Identification of additional (not obvious) locations
- Effectiveness of alternate locations and sizes
- Documentation and consistency of valve locations and sizing

Studies have shown a strong correlation between analysis and system performance (Kroon, 1984).

REFERENCES

AWWA M11. 1989. *Steel Pipe—A Guide for Design and Installation*, 3rd edition, Denver, Colo.

Kroon, Joseph R. et al. 1984. "Water Hammer: Causes and Effects," *Journal AWWA*, November.

Sanks, Robert L., et al. 1989. *Pumping Station* Design, Butterworth-Heinemann, Boston.

Tullis, J. Paul. 1989. *Hydraulics of Pipelines*, John Wiley & Sons, New York.

Wood, D.J., "Surge Reference Manual," Department of Civil Engineering, University of Kentucky, latest edition.

This page intentionally blank.

Chapter 6

Installation, Operation, Maintenance, and Safety

To ensure that the air valve will operate properly, reasonable care is needed in handling, installation, and maintenance. This chapter provides the basic instructions for using air valves, but it is important that the instructions provided with the valve be carefully reviewed and followed.

INSTALLATION

Instruction Manual

The instruction manual supplied by the manufacturer should be reviewed in detail before installing an air valve. At the job site prior to installation, the air valve should be visually inspected and any packing or foreign material in the interior portion of the valve should be removed. The nameplate information on the air valve should be verified to ensure that the valve coincides with that specified.

Location

The air valve should be installed as close to the pipe as possible. The interconnecting piping to the air valve must slope upward toward the valve and be large enough to accommodate the required flow of air. The further the air valve is from the pipeline, the larger the connecting pipe should be.

Shutoff Valve

If a shutoff valve is the same size as the connecting pipe, it should be installed between the air valve and the top of the pipeline to facilitate maintenance (see Figure 6-1). The shutoff valve should be located as close to the main pipeline as possible.

Size of Connection to Pipeline

The size of the connection to the top of the pipeline should equal or exceed that of the air valve inlet connection.

Valve Coating

Internal and external valve corrosion should be controlled by applying the proper coating when necessary.

Bolting Material

All nuts and bolts should be protected to prevent corrosion.

Valves Located Aboveground

Aboveground air valves should be protected from freezing, contamination, or vandalism.

Valves Located Belowground

In addition to the protection from freezing, contamination, and vandalism, air valves located belowground should also be provided with a proper valve vault.

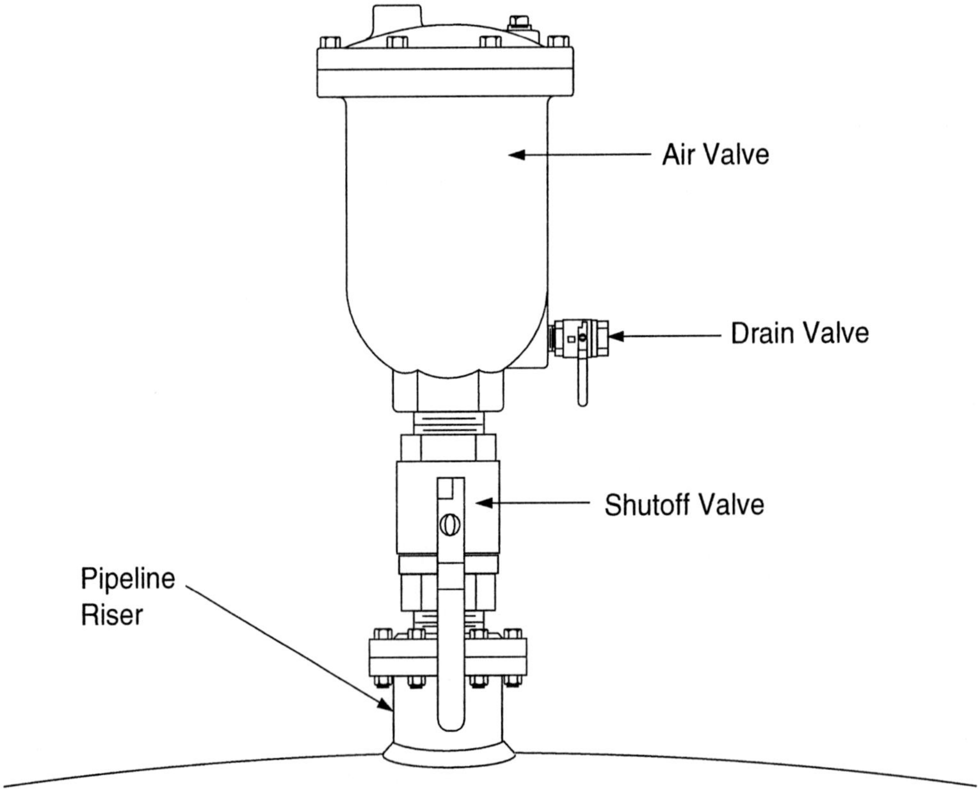

Figure 6-1 Pipeline installation of an air-release valve

Valve Vault

A valve vault should have adequate screened ventilation to satisfy the air requirements for the valve and ventilation of the structure as shown in Figure 6-2. The two vent pipes provide for regular air flow. In freezing conditions, a single vent pipe with baffle can be used. There should be adequate drainage provided to prevent flooding of the vault. Valve vaults should be large enough to provide a minimum of 2 ft (0.6 m) of clearance around and above the air valve for maintenance and valve removal.

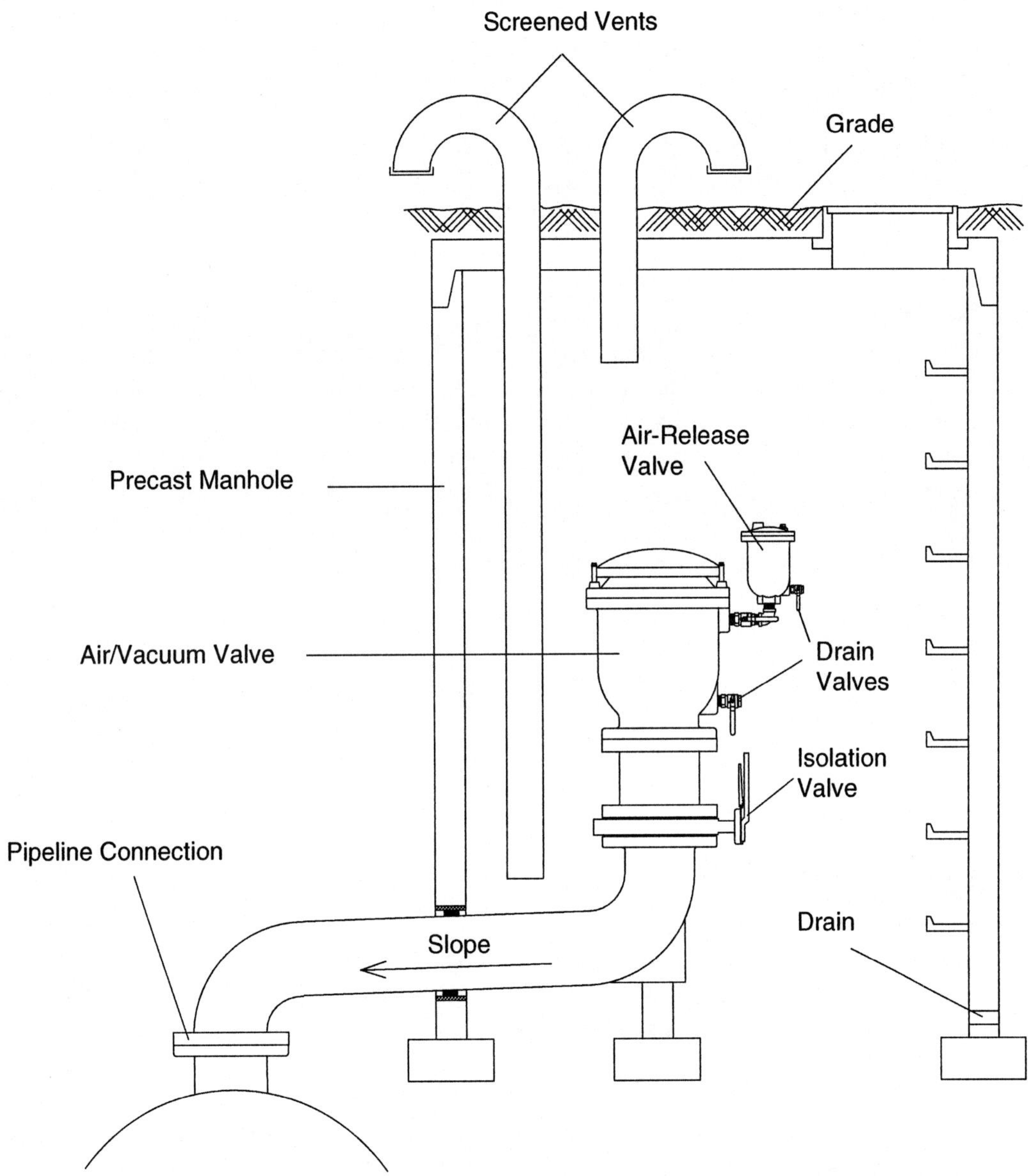

Figure 6-2 Vault installation of a combination air valve

Flooding

Flooding submerges the air intake of air valves, prevents the proper operation of the valve, and may introduce contamination into the pipeline. An outside air intake piped directly to the air valve may help prevent contamination of the pipeline. Provide all intake piping with a down-turned elbow, an air gap, and a bird screen.

Depth of Burial

Valves should be buried below the frost line to prevent freezing. Where combination air valves are used, those in a single body are less likely to freeze than those in separate bodies.

Freezing

Suitable insulation and electrical heat tape should be provided in areas prone to freezing. Frozen air valves will not operate and can be damaged. Thermally activated relief valves (typically supplied in $^3/_8$ or $^1/_2$ in. [9 mm or 13 mm] diameter) can be installed on the valve body to release water and reduce the possibility of damage from freezing. The relief valve automatically opens when the water temperature in the valve falls below a set point (typically 35° F [2°C]) and recloses at higher temperatures.

Contamination

Valves with top-threaded openings should be protected with a protective cap, U-bend, or elbow to prevent rocks, sand, and other particles from falling into the valve. To protect air valves with large metal hoods covering the valve discharge opening from rodents and bird nests, a heavy screened cage covering the air valve outlet should be used.

OPERATION AND MAINTENANCE

The manufacturers' recommendations on air valve operation and maintenance should be followed.

Continuously Operating Air Valves

Air valves that operate continuously should be opened and flushed more often than valves used for filling and draining. All air valves should be opened and flushed at least annually.

Filling and Draining Pipelines

Caution is required when filling or draining pipelines that have air/vacuum or combination air valves installed on the pipeline; see chapter 5. Never prop the valve open by inserting objects into the valve venting port. This can damage the valve seat, and the object may fall into the valve.

Inspection

Air valves should be inspected at least annually for leakage, and the resilient seats should be replaced as necessary.

SAFETY

Underground Structures

Hazardous gases collecting in underground structures have caused injuries and fatalities. Gases drawn into a pipeline can exit through air valves and remain in the underground structure. Always ventilate the underground structure and use a combustible gas and low-oxygen detector before entering the structure. Consult Occupational Safety and Health Administration rules and procedures, such as the need for harnesses and ground-level supervision, in all underground work.

Inspection

When inspecting air valves, isolate the valve by closing the shutoff valve before putting hands and fingers into the valve outlet. If the air valve should suddenly close, hands or fingers could be injured or lost. Pressurized air can also be trapped between the shutoff valve and the air valve; therefore, any removal of air valve bolts, plugs, or covers must be done with extreme care to release trapped air slowly and prevent serious injury.

Pipeline Filling

Thread protectors and packing material should be removed from air valves prior to filling the pipeline.

This page intentionally blank.

Bibliography

AWWA M11. 1989. Steel Pipe—*A Guide for Design and Installation*, 3rd edition. Denver, Colo.

AWWA Standard for Air-Release, Air/Vacuum and Combination Air Valves for Waterworks Service, AWWA C512-92 (1st edition), Denver, Colo.

Colorado State University. 1977. Concepts of Water Hammer & Air Entrapment in the Filling & Testing of Pipelines. Fort Collins, Colo.

Dean, John A. 1992. *Lange's Handbook of Chemistry*. McGraw Hill, New York.

Edmunds, Robert C. 1979. Air Binding in Pipes, *Journal AWWA*, May, 1979.

Flow of Fluids. 1982. Technical Paper No. 410, Crane.

Giles, Ranald V. *Fluid Mechanics and Hydraulics*, (2nd edition) McGraw Hill, New York.

Karassik, Igor J. et al. 1986. *Pump Handbook*, 2nd edition. McGraw Hill, New York.

Kroon, Joseph R. et al. 1984. Water Hammer: Causes and Effects, *Journal AWWA*, November, pp. 39–45.

Landon, P.O. 1994. Air In Pipe? Time to Review Air Valve Basics, *Opflow*, AWWA Vol. 20-3, March.

Lescovich, J.E. 1972. Locating and Sizing Air-Release Valves, *Journal AWWA*, July.

Sanks, Robert L., et al. 1989. *Pumping Station Design*, Butterworth-Heinemann, Boston.

Theory, Application, and Sizing of Air Valves 1997. Val-Matic Valve & Mfg. Corp.

Thorley, A.R.D. 1991. *Fluid Transients in Pipeline Systems*, D.L. George Ltd.

Tullis, J. Paul 1989. *Hydraulics of Pipelines*, John Wiley & Sons, New York.

Wisner, Paul E. 1975. Removal of Air From Water Lines, *Journal of the Hydraulics Division, ASCE*, February.

Wood, D.J. Surge Reference Manual, Department of Civil Engineering, University of Kentucky, latest edition.

This page intentionally blank

Index

This page intentionally blank.

AWWA Manuals

M1, *Principles of Water Rates, Fees, and Charges,* Fifth Edition, 2000, #30001PA
M2 Instrumentation and Control, Third Edition, 2001, #30002PA
M3, *Safety Practices for Water Utilities,* Fifth Edition, 1990, #30003PA
M4, *Water Fluoridation Principles and Practices,* Fourth Edition, 1995, #30004PA
M5, *Water Utility Management Practices,* First Edition, 1980, #30005PA
M6, *Water Meters—Selection, Installation, Testing, and Maintenance,* Fourth Edition, 1999, #30006PA
M7, *Problem Organisms in Water: Identification and Treatment,* Second Edition, 1995, #30007PA
M9, *Concrete Pressure Pipe,* Second Edition, 1995, #30009PA
M11, *Steel Pipe—A Guide for Design and Installation,* Fourth Edition, 1989, #30011PA
M12, *Simplified Procedures for Water Examination,* Second Edition, 1997, #30012PA
M14, *Recommended Practice for Backflow Prevention and Cross-Connection Control,* Second Edition, 1990, #30014PA
M17, *Installation, Field Testing, and Maintenance of Fire Hydrants,* Third Edition, 1989, #30017PA
M19, *Emergency Planning for Water Utility Management,* Fouth Edition, 2001, #30019PA
M20, *Water Chlorination Principles and Practices,* First Edition, 1973, #30020PA
M21, *Groundwater,* Second Edition, 1989, #30021PA
M22, *Sizing Water Service Lines and Meters,* First Edition, 1975, #30022PA
M23, *PVC Pipe—Design and Installation,* First Edition, 1980, #30023PA
M24, *Dual Water Systems,* Second Edition, 1994, #30024PA
M25, *Flexible-Membrane Covers and Linings for Potable-Water Reservoirs,* Third Edition, 2000, #30025PA
M26, *Water Rates and Related Charges,* Second Edition, 1996, #30026PA
M27, *External Corrosion Introduction to Chemistry and Control,* First Edition, 1987, #30027PA
M28, *Cleaning and Lining Water Mains,* Second Edition, 2001, #30028PA
M29, *Water Utility Capital Financing,* Second Edition, 1998, #30029PA
M30, *Precoat Filtration,* Second Edition, 1995, #30030PA
M31, *Distribution System Requirements for Fire Protection,* Third Edition, 1998, #30031PA
M32, *Distribution Network Analysis for Water Utilities,* First Edition, 1989, #30032PA
M33, *Flowmeters in Water Supply,* First Edition, 1989, #30033PA
M34, *Water Rate Structures and Pricing,* Second Edition, 1999, #30034PA
M35, *Revenue Requirements,* First Edition, 1990, #30035PA
M36, *Water Audits and Leak Detection,* Second Edition, 1999, #30036PA
M37, *Operational Control of Coagulation and Filtration Processes,* Second Edition, 2000, #30037PA
M38, *Electrodialysis and Electrodialysis Reversal,* First Edition, 1995, #30038PA
M41, *Ductile-Iron Pipe and Fittings,* First Edition, 1996, #30041PA
M42, *Steel Water-Storage Tanks,* First Edition, 1998, #30042PA
M44, *Distribution Valves: Selection, Installation, Field Testing, and Maintenance,* First Edition, 1996, #30044PA
M45, *Fiberglass Pipe Design,* First Edition, 1996, #30045PA
M46, *Reverse Osmosis and Nanofiltration,* First Edition, 1999, #30046PA
M47, *Construction Contract Administration,* First Edition, 1996, #30047PA
M48, *Waterborne Pathogens,* First Edition, 1999, #30048PA
M49 *Butterfly Valves: Torque, Head Loss, and Cavitation Analysis* First Edition, 2001, #30049PA
M50 *Water Resources Planning,* First Edition, 2001, #30050PA
M51 *Air-release, Air/vacuum and Combination Air Valves,* First Edition, 2001, #30051PA

To order any of these manuals or other AWWA publications, call the Bookstore toll-free at 1-(800)-926-7337.

This page intentionally blank.

Printed in the United States
62382LVS00002B/569-580

9 781583 211526